I0087972

Dark Matter, Dark Energy & the Quantum Wavefunction

Robert Rose

123 Books

Copyright © 2011 by Robert Rose

All rights reserved. This book, or parts thereof, may not be reproduced in any form without permission.

A catalogue record for this book is available from the British Library

ISBN: 978-1-907962-07-3

Published by 123 Books

Reading, England

For Linda

Astrophysicists sometimes argue that all the matter that we have come to know and love in the universe – the stuff of stars, planets, and life – are mere buoys afloat in a vast cosmic ocean of something that looks like nothing.[1]

Atoms and molecules – the stuff of planets and stars – are but minority occupants of space. The major players are bizarre entities known as dark matter and dark energy. Until we fathom these substances, we have taken only a child's step toward comprehending the universe as a whole.[2]

Contents

Preface   11

Introduction   13

1   Dark Matter   17

2   Dark Energy   21

3   The Quantum Wavefunction   24

4   The Big Bang   30

5   Could the Quantum Wavefunction be   38
    Dark Matter?

6   Conclusion   45

Bibliography   47

Preface

Contemporary physics asserts the existence of Dark Matter, Dark Energy and the Quantum Wavefunction. In this book I explore possible links between these three different phenomena and reach a number of thought-provoking conclusions. Firstly, Dark Matter and the Quantum Wavefunction could be one and the same thing. Secondly, both this combined thing and 'ordinary' matter could have evolved out of Dark Energy. Thirdly, that unification of the fundamental forces could include antigravity rather than gravity.

Introduction

When a typical human observes the world that surrounds them they assume that it is a combination of empty space and visible 'material' objects. So, I can observe two trees just outside my window and I assume *both* that these two trees are material objects *and* that the gap between them is empty space. It is seemingly hard to believe that there are 'invisible' material objects, that there is no such thing as empty space, and that visible material objects can cease to exist at a particular place and start existing as potentiality at numerous different places; however, our best physics tells us that all of these beliefs are true.

Physics tells us that 73 percent of the mass of the universe is Dark Energy which exists where we perceive empty space, that 23 percent of the universe is Dark Matter which exists as 'invisible matter', and that the remaining 4 percent of the universe is constituted by 'ordinary' Matter which exists as visible material objects.[3] As if this wasn't hard enough to believe we are also told that the entire universe that we live in was once no bigger than a grapefruit, and that all the Dark Energy, Dark Matter and Matter that constitutes our universe today was once inside this grapefruit-sized area before it exploded in the Big Bang.

Furthermore, quantum physics tells us that visible material objects are constituted out of entities which can change from material particles into a wavefunction of potentialities. When a material entity transforms into a wavefunction it ceases to exist at a particular spatio-temporal location; the wavefunction represents an area of spatio-temporal probability in which the material entity that transformed into the wavefunction is *more or less likely to be found.*

As I have said, most people would find some, if not all, all of these things to be highly dubious, and possibly even ludicrous. However, physics has advanced immensely in the past century and the scientific evidence for the existence of things called Dark Matter, Dark Energy, and the Quantum

Wavefunction is strong. It is not my intention to question the existence of such things; I will assume that they all exist. Rather, my main objective is to reflect on possible links between these three phenomena.

Chapter 1

Dark Matter

The motions of stars within galaxies, and other evidence, tell us that there is a kind of matter out there in addition to the stuff that the stars, the planets, and we are made of. It has gravitational attraction but does not emit, absorb, or reflect light. We thus cannot see it – it's "dark matter." No one knows what it is[4]

Since the 1930s astronomers have been aware of a "missing mass" problem – most of the measured gravitational forces in the universe cannot be explained by observable Matter. This discrepancy between the mass in visible objects and the total mass of systems varies across the universe, ranging from a factor of 2 up to several hundred, and averaging 6.[5] It has been established since the 1980s that the "missing mass" cannot be explained by nonluminous 'ordinary' Matter, so it has been concluded that it must arise from something else. The term Dark Matter has been established to refer to this something else.

It needs to be stressed that the term Dark Matter refers to an aspect of reality about which we are utterly clueless; something is known to exist because it has an identifiable gravitational effect, but that something itself is an unknown. As Brian Greene asserts: "our galaxy and possibly the whole universe is immersed in a bath of *dark matter,* the identity of which has yet to be determined."[6] The term Dark Matter risks causing confusion because it implies the existence of a type of matter that is similar to 'ordinary' Matter except for being nonluminous. The standard position is that Dark Matter is a type of matter but that it is very different to 'ordinary' Matter. However, we "might discover that the dark matter does not consist of matter at all, but of something else."[7]

Clearly, Dark Matter is an elusive phenomenon. What do we know about it apart from the fact that it exists? It has been established that Dark Matter is a *noninteracting substance*. This is because following the Big Bang Dark Matter did not participate in the nuclear fusion that created nuclei. If it did: "then because the dark matter packed six times as many particles into the tiny volumes of the early universe as ordinary matter did, its presence would have dramatically increased the fusion rate of hydrogen."[8] It is generally held that Dark Matter doesn't interact with the strong force, the weak force, or the electromagnetic force; this means it doesn't make nuclei or molecules, or absorb, emit, reflect, or scatter light. It simply exerts gravity to which Matter responds. So, Dark Matter is a very mysterious phenomenon which we know very little about despite it comprising 23 percent of the universe. It isn't nonluminous Matter, but it does exert gravity on Matter, and apart from this influence it is a noninteracting substance. It could itself be a type of matter, or it might not.

Chapter 2

Dark Energy

no one knows what the dark energy is[9]

In 1915 Einstein produced his theory of general relativity which postulated the existence of a 'cosmological constant' – an amount of energy that pervades empty space. Einstein thought that the constant was required because the universe appeared to be static. In the 1920s it was realized that the universe is expanding so the cosmological constant was considered to be unnecessary. However, in 1998 it was discovered that the universe does actually have a non-zero cosmological constant; 'empty' space isn't empty it contains energy. This energy comprises 73 percent of the mass of the universe and has been named Dark Energy. Dark Energy comprises such a large amount of the universe because most of the universe is 'empty space' and the energy in this space is transferrable into mass through $E = mc^2$.

The evidence for the existence of Dark Energy came from the observation of supernovae; these exploding stars gave off light on explosion that was fainter than expected. This was due to the universe expanding more rapidly than it would have if space was empty; the excess speed of universal expansion was due to the presence of the Dark Energy. Whilst matter slows the expansion of the universe through gravity pulling everything towards everything else, Dark Energy makes space expand and thereby accelerates the expansion of space.

Dark Energy must be an exceptionally powerful force to be able to push all the matter of the universe apart. However, as with Dark Matter, Dark Energy is a very mysterious entity which is known to exist, but about which very little else is known. Given that it constitutes 73 percent of the mass of the universe our ignorance of the nature of Dark Energy necessarily entails ignorance about the nature of the vast majority of reality.

Chapter 3

The Quantum Wavefunction

Quantum probability is not the probability of where the atom is. It's the objective probability of where you (or anyone) will find it. The atom wasn't in that box before you observed it to be there. Quantum theory has the atom's wavefunction occupying both boxes. Since the wavefunction is synonymous with the atom itself, the atom is simultaneously in both boxes.[10]

The modern foundations of quantum mechanics were laid down by Erwin Schrodinger in the 1920s. According to quantum mechanics a particle – such as an atom – can exist either as a concentrated particle or as a spread-out wave. When an atom is in its wave state it forms a wavefunction of potentialities in which it ceases to exist at a particular location; the Schrodinger Equation describes how these wavefunctions change over time. This picture seems hard to accept because whenever we observe an atom *it is* in a particular location. However, experiments consistently seem to show that when we are not observing an atom *it isn't* in a particular location; it is in a "superposition" state of potentiality. So, the wavefunction represents an area of spatio-temporal probability in which the atom that transformed into the wavefunction is more or less likely to be *found*; the atom isn't in a particular place *until* it is found.

The classic experiment is called the double-slit experiment. In the experiment a beam of light is shone onto a barrier with two vertical slits which is placed in front of a photographic plate. When only one of the two slits is open a single vertical line appears on the photographic plate directly behind the open slit. However, when both slits are open an interference pattern appears. Even when photons are fired one at a time towards the two open slits an interference pattern still emerges. This means that the photon must go through both slits simultaneously and interfere with itself on the other side; it must be a wave. Furthermore, if an attempt is made to observe which of the two slits a photon passes through then no interference pattern occurs; observation collapses the wavefunction into a particle. Therefore, a photon can exist either as a particle in a particular place, or as a wavefunction in different places at the same time. The same effects are observed with electrons and atoms.

The quote at the start of this chapter is the standard interpretation of quantum probability and how it applies to the Quantum Wavefunction. The claim that 'the atom is simultaneously in both boxes' is so contradictory to common sense that many people will assert that it must be false. However, the claim is worded in such a way that this assertion is understandable. The wavefunction *is* synonymous with the atom, but this surely doesn't entail that *the atom* is in multiple places at the same time. In its wavefunction state *the atom doesn't exist*, all that exists in multiple places is the *potentiality* for the atom to return into a particle form. If something doesn't exist then it *cannot* be simultaneously in two places. When we assert that it is 'the potentiality to become an atom that is simultaneously in both boxes' then quantum probability doesn't collide with common sense in such a violent manner.

However, Quantum Wavefunctions are still deeply mysterious things as they imply an observer created reality. The state of photons, electrons, and atoms is dependent on whether they are currently being observed. If they are observed they are particles located in a particular place; if they are not observed they are a wavefunction of potentialities that is located in multiple places. These microscopic particles are the constituents of all Matter; this means that all Matter can change from a particle into a spread-out wavefunction of potentialities.

Chapter 4

The Big Bang

Having looked at Dark Matter, Dark Energy and the Quantum Wavefunction as separate phenomena it is now time to look at possible links between them. Our first area of possible linkage is the Big Bang. The mass of the universe today is 73 percent Dark Energy, 23 percent Dark Matter, and 4 percent Matter. Were all of these present at the Big Bang? There has to be a serious possibility that they were not. We know from $E=mc^2$ that matter can change into energy at the appropriate temperature. This means that Dark Energy is transferrable into Matter. Furthermore, whatever Dark Matter is it too will presumably be transferrable into Matter or Dark Energy. If all three of our phenomena are transferrable into each other then it has to be possible that at the Big Bang they were all part of a single substance.

What could this one substance be? The obvious candidate is Dark Energy as it constitutes 73 percent of our universe today. And, of course, the hallmark of Dark Energy is that it has an antigravitational force which *causes* the universe to expand. As the initiator of an expanding universe, and the possessor of such immense power that today it forces the universe to expand in spite of gravity, it is surely the

31

case that in the past this power would have been much more intense as it was spread over a vastly smaller universe. Indeed, at the moment of the Big Bang the amount of *'expanding universe energy'* bottled up in the Dark Energy of the grapefruit-sized area of space would have been mind-blowing. It seems that Dark Energy could have been the initiator of the Big Bang, the force behind its immense power, and the only substance in existence at the time.

Shortly after the Big Bang Dark Energy could have given rise to Dark Matter, which is a sizeable 23 percent of the universe today, and shortly after this event Dark Matter in turn could have given rise to Matter. Alternatively, it could have taken a *long time* for the Dark Matter to Matter transition to occur. It is entirely possible that the transformation from Dark Energy to Dark Matter requires an immensely high temperature, whilst the transformation from Dark Matter to Matter requires a much lower temperature; if this is so, then the Dark Matter-Matter transformation could have happened a long time after the Big Bang.

This sequence of transformations fits the respective masses of the various phenomena perfectly. The reason Dark Energy has 73 percent of the mass of the universe today would be because it is the primordial and fundamental underlying state of the universe. It would have had 100 percent of the universal mass at the Big Bang before it transferred some of its energy into the formation of Dark Matter – a sizeable 23 percent of the universe. In turn, Dark Matter transformed into Matter which constitutes a paltry 4 percent of the universe today.

Some physicists will, no doubt, consider such speculation to be premature. We have seen that the nature of both Dark Matter and Dark Energy *is* a complete mystery, *but* physicists typically assume that they are very different things. Furthermore, Dark Matter is typically assumed to be a kind of matter and therefore not something that could give rise to Matter. However, whilst the nature of Dark Matter continues to be a mystery, the idea that Dark Energy gave rise to Dark Matter, which gave rise to Matter, has to be considered.

Indeed, in the next section I will propose that the 'mysterious' Dark Matter is, in fact, simply the Quantum Wavefunction. If this is so, then the Big Bang transformations I have been describing become much more intelligible. Dark Energy will be *pure energy*, which at the moment of the Big Bang transferred some of this energy into Dark Matter, which we can now conceive of as the '*wavy energy of potentialities*' of the Quantum Wavefunction. Almost instantaneously, or at a much later time, some of this '*wavy energy of potentialities*' exercised some of its potential to give rise to Matter.

In this scenario we can see that antigravity is the normal underlying state of the universe. It is only when Dark Matter emerges from Dark Energy that a gravitational force emerges; gravity is an attribute of Dark Matter which slows down the antigravitational force of Dark Energy. Matter, as something that emerged from Dark Matter, shares its gravitational properties. Physicists are puzzled as to why gravity is so immensely weak compared to all of the other fundamental forces. If the scenario outlined here is correct we can immediately see why gravity is immensely weaker than the other forces; it is actually a 'negative' force, the result of the weakening of the 'positive' force of antigravity. At the Big Bang gravity didn't exist, what existed was antigravity. It is antigravity, not gravity, which will have strength comparable to the other forces. A unification of the fundamental forces would therefore include the strong nuclear force, the weak nuclear force, electromagnetism and antigravity.

In this scenario it further follows that when people are talking about wave-particle duality, they are simultaneously talking about Dark Matter-Matter duality. The more fundamental state is the Quantum Wavefunction / *'wavy energy of potentialities'* of Dark Matter. Whilst the particle state of Matter is less fundamental.

The scenario I have outlined might seem to be fairly intelligible. However, its intelligibility depends on the mysterious Dark Matter actually being the Quantum Wavefunction. It is now time to explore whether this could be the case.

Chapter 5

Could the Quantum Wavefunction be Dark Matter?

We have seen that both Quantum Wavefunctions and Dark Matter are extremely mysterious things. I have also argued that the universe could have originated with one substance – Dark Energy – which gave rise to Dark Matter, and that in turn Dark Matter gave rise to Matter. If this is what actually happened then there has to be a possibility that these transformations are still occurring today. Could it be that we have actually already unwittingly 'observed' some of these transformations via experiments on the quantum wavefunction?

There are obvious similarities between Dark Matter and the Quantum Wavefunction. Neither has been directly observed, both have been postulated to exist as scientific necessities, both are considered to be 'invisible', and both are possibly unobservable in principle. The Quantum Wavefunction itself is an area of potentiality which can transform into a particular particle. It is surely possible that the 23 percent of the mass of the universe that is Dark Matter is a massive 'sea of potentialities', an aggregation of individual wavefunctions, some of which have previously been Matter. Interestingly:

> *In 2007 astronomers released the first large-*
>
> *scale three-dimensional map of dark matter*
>
> *in the universe. The map provides the best*
>
> *evidence yet that normal matter – stars and*
>
> *galaxies – collects within the densest concen-*
>
> *trations of dark matter.*[11]

In other words, wherever there is Matter there is also Dark Matter, this would be exactly what we would expect to be the case if Dark Matter is Quantum Wavefunctions of Matter. We would also expect there to be places where there is Dark Matter and no Matter – as not all of the potentialities for transformation will be actualized. So, it is also of interest that in 2004 VIRGOHI21 was discovered: "the first known completely dark galaxy".[12] This picture fits very nicely with our knowledge that Dark Matter is a *noninteracting substance*; it doesn't *interact with* Matter – it *turns into* Matter.

It would follow that all Matter is 'surrounded' by the Dark Matter that gave rise to it, and that in certain conditions Matter can transform back into its surrounding Dark Matter, into a more 'primitive' state of potentiality. This would perfectly explain what occurs in quantum interference experiments, and fits perfectly with the Big Bang scenario outlined in the previous section. However, this might all seem a bit too speculative. Are there any existing theories which could give support to this picture by establishing a link between Dark Matter and the

Quantum Wavefunction? In fact, there are, and they are called Supersymmetric Theories.

Supersymmetric Theories attempt to give a common account of the two types of particles: fermions and bosons. These theories predict that every standard elementary particle will have a supersymmetric counterpart which has an opposite spin. These hypothesized entities are referred to as sparticles, but none of them have ever been found. It is entirely possible that sparticles could be Dark Matter. Furthermore, and very importantly, one particular sparticle – the neutralino – is the super-symmetric counterpart of *several different* bosons:

It [the neutralino] forms an amalgamated companion of the photon (photino), Z boson (zino), and Higgs boson (higgsino), mixed together in a quantum state.[13]

So, if Supersymmetric Theories are correct their hypothesized sparticles could be the *'wavy energy of potentialities'* that are simultaneously Quantum Wavefunctions and Dark Matter. If this is so, it would explain why sparticles have never been found – any attempt to find them collapses their wavefunctions into their supersymmetric counterpart particles of Matter. The hypothesized sparticles clearly give us a framework for understanding how Matter can transform into a Quantum Wavefunction of potentialities. Furthermore, if sparticles such as the neutralino are Dark Matter then we also have the tools for understanding how Quantum Wavefunctions could be Dark Matter. And, according to Halpern and Wesson: "the neutralino fits the profile of a dark matter candidate well".[14]

Chapter 6

Conclusion

I have explored the links between three phenomena that contemporary physics asserts to exist – Dark Matter, Dark Energy and the Quantum Wavefunction. These three phenomena are usually considered to be different things, but we have seen that there are good grounds for believing that Dark Matter and the Quantum Wavefunction could be one and the same thing. We have also seen that Matter could have emerged from Dark Matter, and that Dark Matter could have emerged from Dark Energy. Furthermore, we have seen that unification of the fundamental forces could include antigravity rather than gravity.

Bibliography

[1] Tyson, Neil deGrasse, and Goldsmith, Donald, *Origins: Fourteen Billion Years of Cosmic Evolution*, (London: W. M. Norton & Company, 2005), p. 70.

[2] Halpern, Paul, and Wesson, Paul, *Brave New Universe*, (Washington: Joseph Henry Press, 2007), p. 21.

[3] Halpern and Wesson, *Brave New Universe*, p. 25.

[4] Rosenblum, Bruce, and Kuttner, Fred, *Quantum Enigma*, (London: Gerald Duckworth & Co. Ltd, 2007), p. 196.

[5] Tyson and Goldsmith, *Origins*, p. 68.

[6] Greene, Brian, *The Elegant Universe*, (London: Vintage, 2000), p. 225.

[7] Tyson and Goldsmith, *Origins*, p. 74.

[8] Tyson and Goldsmith, *Origins*, pp. 72-3.

[9] Rosenblum and Kuttner, *Quantum Enigma*, p. 196.

[10] Rosenblum and Kuttner, *Quantum Enigma*, pp. 102-3.

[11] Baumann, Mary K., and Hopkins, Will, and Nolletti, Loralee, and Soluri, Michael, *Cosmos*, (London: Duncan Baird Publishers, 2007), p. 286.

[12] Halpern and Wesson, *Brave New Universe*, p. 122.

[13] Halpern and Wesson, *Brave New Universe*, p. 141.

[14] Halpern and Wesson, *Brave New Universe*, p. 141.

www.ingramcontent.com/pod-product-compliance
Lightning Source LLC
Chambersburg PA
CBHW071438040426
42445CB00012BA/1386